"No one but N
links the twin,
nized human existence – cataclysmic climate change and nuclear doomsday machines – and no previous communications of his warnings and challenge to action has presented them so impressively."

—*Daniel Ellsberg, the* Pentagon Papers *whistleblower*

Internationalism or Extinction

In his new book, Noam Chomsky writes cogently about the threats to planetary survival that are of growing alarm today. The prospect of human extinction emerged after World War II, the dawn of a new era scientists now term the Anthropocene. Chomsky uniquely traces the duality of existential threats from nuclear weapons and from climate change – including how the concerns emerged and evolved, and how the threats can interact with one another. The introduction and accompanying interviews place these threats in a framework of unprecedented corporate global power, which has overtaken nation states' ability to control the future and preserve the planet. Chomsky argues for the urgency of international climate and arms agreements, showing how global popular movements are mobilizing to force governments to meet this unprecedented challenge to civilization's survival.

Considered the founder of modern linguistics, **Noam Chomsky** is one of the most cited scholars in modern history and among the few most influential public intellectuals in the world. He has written more than 100 books, his most recent being *Requiem for the American Dream: The 10 Principles of Concentration of Wealth & Power.* Before coming to the University of Arizona as Laureate Professor of Linguistics in 2017, Chomsky taught at Massachusetts Institute of Technology for 50 years.

Charles Derber is Professor of Sociology at Boston College.

Suren Moodliar is managing editor of *Socialism and Democracy.*

Paul Shannon is program staff for the Peace and Economic Security program of the American Friends Service Committee (AFSC).

ROUTLEDGE UNIVERSALIZING RESISTANCE SERIES

Co-Edited by Charles Derber and Suren Moodliar

The modern social sciences began in the late 19th century when capitalism was establishing itself as the dominant global system. Social science began as a terrifying awakening: that a militarized, globalizing capitalism was creating the greatest revolution in history, penetrating every part of society with the passions of self-interest and profit and breaking down community and the common good. The universalizing of the market promised universal prosperity but delivered an intertwined sociopathic system of money-making, militarism and environmental destruction now threatening the survival of all life itself.

In the 21st century, only a universalized resistance to this now fully universalized matrix of money, militarism and me-firstism can save humanity. History shows that people can join together under nearly impossible odds to create movements against tyranny for the common good. But when the world faces a universalizing system of madness and extinction, it takes new forms of resistance moving beyond the "silo" movements for social justice that have emerged notably in the US in recent decades: single-issue movements separated by issue, race, gender, social class, nation and geography. The story of what universalized movements look like, how they are beginning to be organized, how they "intersect" with each other against the reigning system of power, and how they can grow fast enough to save humanity is the purpose of this series.

The series is publishing works by leading thinkers and activists developing the theory and practice of universalizing resistance. The books are written to engage professors, students, activists and organizers, and citizens who recognize the desperate urgency of a universalizing resistance that can mobilize the general population to build a new global society preserving life with justice.

Published Books

Charles Derber, *Welcome to the Revolution: Universalizing Resistance for Social Justice and Democracy in Perilous Times* (2017)

Charles Derber, *Moving Beyond Fear: Upending Security Tales in Capitalism, Fascism and Democracy* (2019)

Yale Magrass and Charles Derber, *Glorious Causes: Irrationality in Capitalism, War and Politics* (2019)

Noam Chomsky, *Internationalism or Extinction* (2020)

Planned

Noam Chomsky, *Chomsky for Activists*

Elisa Batista and Matt Nelson, *¡Presente! Latinx Power Remaking Democracy*

Chuck Collins, *Disrupting Narratives of Deservedness: Changing the Stories that Justify Economic and Racial Inequality*

Charles Derber and Suren Moodliar, *After Midnight: How Modern-Day Abolitionism Can Save the Planet*

Suren Moodliar, *Revolution Has an Address! The Transformative Power of Movement Building Spaces*

Internationalism or Extinction

Noam Chomsky

Edited by
Charles Derber,
Suren Moodliar, and
Paul Shannon

NEW YORK AND LONDON

First published 2020
by Routledge
52 Vanderbilt Avenue, New York, NY 10017

and by Routledge
2 Park Square, Milton Park, Abingdon, Oxon, OX14 4RN

Routledge is an imprint of the Taylor & Francis Group, an informa business

Library of Congress Cataloging-in-Publication Data
A catalog record for this title has been requested

ISBN: 978-0-367-43061-0 (hbk)
ISBN: 978-0-367-43058-0 (pbk)
ISBN: 978-1-003-00103-4 (ebk)

Typeset in Bembo
by codeMantra

Contents

Introduction 1

1 The Twin Threats 14

2 Reaching People 46

3 Thinking Strategically 62

4 Updated Reflections on Movements 75

5 The Third Threat: The Hollowing Out of Democracy 86

6 To Learn More 99

Index 103

Introduction

The urgency of "looming extinction" cannot be overlooked. It should be a constant focus of programs of education, organization, and activism, and in the background of engagement in all other struggles. But it cannot displace these other concerns, in part because of the critical significance of many other struggles, and also in part because the existential issues cannot be addressed effectively unless there is general awareness and understanding of their urgency. Such awareness and understanding presupposes a much broader sensitivity towards the tribulations and injustices that plague the world – a deeper consciousness that can inspire activism and dedication, deeper insight into their roots and linkages. There's no point calling for militancy when the population is not ready for it, and that readiness has to be created by patient work. That may be frustrating when we consider the urgency of the existential threats, which is very real. But frustrating or not, these preliminary stages cannot be skipped.

– Noam Chomsky, December 2018

On a hazy mid-October afternoon, in 2016, just before the fateful November election that would thrust Donald J. Trump into the White House, a large crowd began to gather outside Boston's historic Old South Church. Soon the crowd would reach beyond two blocks. Though all concerned were highly animated by the coming elections, voting was not the only thing on their minds. Some having traveled across national borders, they were there to attend a "Chomsky event" – the generic title for the distinctive, large public lecture and conversation that plays out countless times when the distinguished linguist and public intellectual agrees to address an audience. Indeed, the youthful attendees lining the sidewalk were about to engage with Noam Chomsky much as their grandparents would have some 50 years earlier when he helped launch the public questioning of a still escalating US intervention in Vietnam. Drawing on widely available sources, Chomsky would structure a presentation, composed in eloquent but spare prose, readily accessible in argument and vocabulary, to the vast majority of the audience. If this Chomsky event followed past scripts, it would undoubtedly provide for a lengthy question and answer period in which Noam would field questions, comments and, infrequently, even heckles from the attendees. Each response would likely receive the same calm, considered attention as the main talk itself. The only likely exception would be topics that call on Noam to speak about *himself*. These would be treated differently, ignored, brushed past, and sometimes deftly dismissed. The deeply egalitarian and democratic Chomsky seems to find such questions irrelevant. The facts and arguments that he marshals in service of "people's causes" render such questions extravagances.

And, the cause to be advanced in October 2016 was a bit different from many Chomsky had addressed in recent years. Not concerned that evening with this or that atrocity or transgression by a superpower, Chomsky had entitled his lecture, "Internationalism or Extinction." The latter referred not to any particular foreign or domestic policy or disaster but the prospects of destroying nearly *all* species of life on the planet.

Chomsky's assembling audience chatted patiently and quietly as they whittled away the hour before the doors finally opened. Certainly, the lecture's title was a plain warning of the apocalyptic topic to be addressed. But what is there to prepare even an educated audience for the fact that their lecturer was about to have them consider the potentially terminal events of most species, including their own? The waiting audience's experience is surely that of the present reader confronting a slim book advertised by as intimidating a title as any offered a modern readership.

But that is the promise of Noam Chomsky: difficult facts and imposing social structures are all susceptible to human reason. Calm deliberation, exchanges of perspectives, clearly formulated arguments and concepts, unadorned historical narratives, strategic questioning, and a collective engagement to persuade and/or pressure and/or overcome the sources of the destruction are all part of the unstated, implicit *activist* commitment of a Chomsky lecture.

This book follows the Chomsky event. The main body, Chapter 1, consists of the original lecture, supplemented by editorial notes that point the reader to additional resources and source materials. It is followed, in Chapter 2, by the transcript of a conversation at the event with Wallace Shawn, a committed activist but someone who is best known as an accomplished dramatist and actor. Building on a friendship initiated in Sandinista Nicaragua of the 1980s, Wally Shawn reflected on Noam's lecture and called on him to address the ever-challenging question – *how do we convince the people who were not in the room to care, to take action?*

The Chomsky answer to this question must have felt unsatisfying to the audience and perhaps even to Shawn himself. They were treated to an account of the various opportunities for treaty making as well as useful historical precedents and rationales for such treaties. Rather than representing a dismissive attitude to the question, Wallace Shawn and the audience were presented with what seems to be as close to a "Chomsky dogma" as anything uttered by the distinguished thinker: *we convince people to care and to act by presenting them with the facts and opportunities. These treaties were opportunities.* Whether his listeners choose appropriate courses of action, however, is not something guaranteed. Implicitly, *history is in our hands, our creativity... and our limits.*

Through the audience discussion period – the transcript of which forms Chapter 3 – following the Wallace Shawn conversation, as with every Chomsky event, variations on this question and answer would repeat themselves. Although the underlying answer never changes, each response is rich in detail and carefully argued, respecting the historical specificity of each topic and therefore the distinctive dilemmas of those wishing to intervene with respect to that topic. No struggle, however local and particular, is given short shrift. The challenge for would-be change makers, then, is how to articulate specific struggles with general ones, especially those facing humanity as a whole

Noam's direct answer, implied by the respect he has for local struggles, is *explicitly* addressed in Noam's postscript to the talk – Chapter 4 – and consists of notes written by Noam in 2019 to update his analysis for the post-election period and the first two years of the Trump administration. As signaled with our opening quote, the threat of extinction does not negate other struggles that may have a more immediate character. Nonetheless, these have to be understood in their relation to the broader, universal struggle for survival *with* justice. People are not expected to surrender either immediate needs or long-developing historical claims; instead, these are to be articulated and intertwined with the struggle against extinction. A final substantive section – Chapter 5 – consists of a new speech carefully drafted by Noam to expand on a *third* existential threat – the undermining of democracy, which, in turn, exacerbates the climate change and nuclear threats.

But what of the main lecture – Internationalism or Extinction – itself? Building on his long-standing opposition to nuclear weapons, Noam introduces his audience to still another threat to "the 200,000-year-old human experiment": climate change. He notices the coincidence between the two threats, with both emerging after World War II (1939–1945). In the months preceding the talk, a working group within the International Union of Geological Sciences had proposed the concept of the "Anthropocene" to indicate that humanity and its social systems have become actual forces of nature – restructuring the planet at the geomorphological level.

Once an obscure concept used by Soviet scientists to suggest the long-term impacts of humanity as a force of nature, the Anthropocene has wended its way through academic discourse and into the mass media as the successor epoch to the Holocene, which began about 11,000 years ago. Carbon levels in the atmosphere, now radically higher than at any previous point in human history, constitute a distinct, objective measure of this impact; human activities, primarily the burning of fossil fuels, drive this still accelerating index. In the lecture, Chomsky demonstrates that this history is intertwined with the once parallel threat of terminal nuclear conflict. Within the Anthropocene epoch, scientists have noted the period of "Great Acceleration," in which carbon concentration levels began their rapid rise to more than 400 parts per million (ppm), significantly higher than the 350 ppm thought to be a safe level. The acceleration began around 1950.[1]

Among environmental scientists and public intellectuals, there is some debate that "Anthropocene" as a designation provides little insight into the social systems that drive these extinction threats. In fact, one important commentator, the environmental historian Jason Moore, believes that we ought to define the epoch that begins in the late 18th century, the "Capitalistocene," to better indicate the *causes* of the epoch's destructive character.

While Chomsky does not address this matter in the present work, he nonetheless considers two elements of human agency that speak to the matter. In one case, he asks his audience to "ponder on the most extraordinary fact: a major political organization in the most powerful country in the world's history is quite literally dedicated to the destruction of much of life on Earth." Here he calls our attention to the Republican Party and both its organized denialism and pro-actively destructive environmental policies. Chomsky's audience may well have questions about the forces that have shaped the Republican Party and the larger system.

With the second case, Chomsky offers an oblique but suggestive answer. He quotes James Madison, a "founding father" and the

4th president of the United States, about the "daring depravity of the times," in which "stock jobbers" (capital-rich speculators) fuse their power with government's, becoming "at once its tools and its tyrant," overwhelming popular rule "by clamors and combinations." In other words, going back to the earliest days of the American republic, private interests captured the state and crowded out popular power and interests in favor of their own logic of profit maximization.

Against these private interests, which Chomsky explores in other works (see Chapter 6: To Learn More), and related "national" interests of the United States, the audience had an opportunity to explore the ways in which international cooperation emerges from both elite and democratic pressures. And yet, as Chomsky's narrative suggests, even these were insufficient for protecting humanity and the planet from the threat of nuclear holocaust. He cites two instances where provocative actions by the United States could have led to an uncontrolled escalation into full-scale nuclear war. Treaties, institutional mechanisms, did not protect us in the two examples offered. During both the so-called "Cuban" missile crisis of the 1960s and "Operation Able Archer" of the 1980s, decisions by field officers to violate protocols and *not* report threatening actions to their superiors allowed humanity to live another day. In the case of Able Archer, Stanislav Petrov didn't report the information to a superior (violating protocol, saving us from likely destruction). During the missile crisis, Vasily Arkhipov refused to authorize the launching of nuclear-tipped missiles. Protocols worked in this case, but barely. Two other officers had agreed on the launch, fortunately protocol required three officers to agree.

If these officers had unreflexively followed standard operating procedures, neither Chomsky nor his audience would be alive to reflect on the actions of these still relatively unknown military officers.

In valorizing this individual-level resistance, Chomsky successfully conveys the tenuous character of our survival and the need for remaking the international order. Although he finds hope in some of the more rational and enlightened elites and their projects, e.g. calls from people like George Schultz (Secretary of State under Reagan) for a nuclear weapons freeze and eventual abolition, he is very conscious that we "cannot expect systems of organized power … to take appropriate actions … unless they are compelled to do so by constant dedicated popular mobilization and activism." As a result, he approvingly singles out as both exemplary and necessary the "huge popular mobilizations" of the early 1980s that opposed nuclear weapons development.

During the question and answer period, Noam provided some insights into his personal engagements. In one anecdote, about the weapons research facility, the Draper Laboratory, he explained the strategic logic informing his perspective. Liberals who opposed Pentagon-funded research at MIT demanded that such activities not be permitted *on campus.* Conservatives, by contrast, were just fine with these happening on campus. The "radical" position, with which Noam identifies, was that should such activities take place, it was preferable that they happen *on campus* where they could be subjected to public scrutiny and debate. The defect of the liberal position, from his point of view, was that it did not eliminate such research; it only relocated it to places beyond the organizing reach of campus-based resistance. In the same vein, thinking about what activists should demand both at the grassroots level and also in concert with or even against state actors informs Noam's presentation in a careful blend of grounded, pragmatic strategies and visionary ambitions.

As Noam's audience began filing into the huge church sanctuary hosting the event, they worked their way through a chapel filled with tables and stalls staffed by many organizations, each addressing topics believed to be of shared concern with the Chomsky-event participants. In other words, if Noam's talk promised a big-picture synthesis of many concerns, piggybacking onto this framework were many very specific struggles. Not elaborated in this book, but opening its companion video, *Noam Chomsky – Internationalism or Extinction* (available for free streaming from ChomskySpeaks.org), are examples of the diverse organizations rallying around the "Noam Chomsky event." These include groups in solidarity with Haiti and Venezuela, local chapters of anti-nuclear and peace organizations, corporate accountability movements, environmental projects, and socialist organizations. The event, the video, and the production of this book were made possible by a grant from the Wallace Action Fund. Its founder, Randall Wallace, is a long-time and careful Chomsky reader. Coincidently, he is the grandson of Henry Wallace, Franklin Delano Roosevelt's first vice president – an agronomist and ecological thinker whose 1948 candidacy for president warned of the emerging Cold War and its foreseeable consequences – ones that Noam so ably depicts in this book.

Against the bleak analysis, even one balanced as this is by a grounded optimism in resistance from below, readers must wonder, as one of Noam's interlocutors did from the floor of that 2016 event, "how do we keep up our spirits?" Noam's characteristically terse reply was simply, "what is the alternative?" "No surrender!" was the unarticulated but widely felt conclusion in the minds of most of the audience, as one must imagine is true for his readers as they respond to the following call for an emergency mobilization against the ongoing sixth mass extinction. In Noam's words: "The tasks ahead are daunting, and they cannot be deferred."

Note

1 By March 2019, the Scripps Institution of Oceanography in San Diego reported Carbon levels at 411.97 ppm, up from March 2017's 407.06 ppm, see https://www.co2.earth/. This indicates that far from limiting emissions, the world remains on track to irreversible, abrupt climate change.

1 The Twin Threats

Extinction and internationalism have been linked in a fateful embrace ever since the moment when the threat of extinction became an all too realistic concern, August 6th, 1945. It's a day that will never be forgotten by those who were alive then and had their eyes open – a day I personally remember very well. On that day, we learned that human intelligence had devised the means to bring the human experiment of 200,000 years to an end.

It was never seriously in doubt from the first moment that the capacity to destroy would escalate and it would diffuse to other hands – increasing the threat of self-annihilation. In the years that followed, the record of near misses is appalling, sometimes from accident and error, a few shocking cases from recklessness, and the threat is growing ominously. A review of the record reveals clearly that escape from catastrophe for 70 years has been a near miracle and such miracles cannot be trusted to perpetuate themselves.

On that grim day in August 1945, humanity entered into a new era, the Nuclear Age. It's one that's unlikely to last long: *either we will bring it to an end, or it's likely to bring us to an end.* It was evident at once that any hope of containing the demon would require international cooperation. By the fall of 1945, a book calling for world federal government reached the top of the bestseller list – it was written by Winston Churchill's literary agent, Emery Reves.[1]

Albert Einstein was only one of those who reacted by calling for *a world government* – that is what he called the political answer to the shattering events of August 1945. They recognized these to be a turning point in human history and, perhaps, the opening of its final stage. Hopes that the United Nations might begin to fulfill [the world government] function were quickly dashed – in itself an important topic but one that I do not explore here.

It was not understood at the time, but a second and no less critical new era was beginning at the same time – a new geological epoch called the Anthropocene. It's an epoch defined by extreme human impact on the environment. Now it's understood that we are well into this new epoch, but there have been disagreements among scientists about just when change became so extreme as to signal the onset of the Anthropocene. In April 2016, the Anthropocene Working Group, an official geological organization, reached a conclusion on the onset of the epoch. They recommended to the 35th International Geological Congress that the dawn of the Anthropocene should be given the time frame that begins after the end of World War II.[2]

According to their analysis then, the Anthropocene and the Nuclear Age coincide; it's a dual threat to the perpetuation of organized human life. Both threats are severe and imminent. It's widely recognized that we have entered the period of the sixth mass extinction. The fifth extinction, 66 million years ago, is generally attributed to an asteroid – a huge asteroid that hit the Earth destroying 75% of the species on Earth. It ended the age of the dinosaurs and opened a way for the rise of small mammals – and ultimately humans about 200,000 years ago.

It hasn't taken us long to bring about the sixth extinction, which is expected to be similar in scale to the earlier ones, though differing in an instructive way. In the mass extinctions that long predate humans, body size was not correlated with extinction. It was kind of an equal opportunity killer – independent of your body size. In the human-generated sixth extinction, which is now underway, the larger animals are being killed disproportionately.

That actually extends a record that traces back to our early proto-human ancestors. They were a predatory species that caused significant harm to large organisms, wiped many of them out, with themselves not too far in the distance. The human capacity to destroy one another on a massive scale has not been in doubt for a long time, and it reached a hideous peak in the past century. The Anthropocene Working Group reaffirms the conclusion that climate-warming CO_2 emissions are increasing in the atmosphere at the fastest rate in 66 million years.

They cite a report last July (2016) that particles of CO_2 reached over 400 parts per million (ppm) and that the level is rising at a rate unprecedented in the geological record. Subsequent studies have revealed that figure was not a fluctuation. It appears to be permanent, a base for further growth, and that figure, the 400 ppm, has been regarded as a critical danger point. It's perilously close to the estimated level of stability of the huge Antarctic ice sheet. Collapse of the ice sheet would have catastrophic consequences for sea level, and these processes are already underway quite ominously in the Arctic regions.

The Rim Fire in the Stanislaus National Forest in California, August 17, 2013.
Source: United States Department of Agriculture (public domain).

The broader picture is no less ominous and practically every month breaks new temperature records; huge droughts are threatening survival for hundreds of millions of people. These are also factors in some of the most horrendous conflicts, Darfur and now Syria. Some 31.5 million people are displaced by disasters such as floods and storms every year; that's a predicted effect of global warming, that's almost one person, every second. These are considerably more than those fleeing from war and terror. The numbers are bound to increase as glaciers melt and sea levels rise, threatening water supplies for vast numbers of people.

The melting of Himalayan glaciers may eliminate the water supply for South Asia, several billion people. In Bangladesh alone, tens of millions are expected to flee in coming decades simply because of sea-level rise; it's a flat coastal plain. That's a refugee crisis that will make today's pale into insignificance and it's barely the beginning. With some justice, Bangladesh's leading climate scientists said recently that these migrants should have the right to move to countries from where all these greenhouse gasses are coming – that millions should be able to go to the United States, which raises a moral issue here of no slight triviality.

Puerto Rico, the morning after Hurricane Maria, September 19, 2017.
Photo: Roosevelt Skerrit (public domain).

Well I won't take time to review the larger record; I presume most of you are pretty familiar with it, but it should be deeply alarming and to anyone concerned with the fate of the species and the other species that we are destroying with abandon. It's not in the far future; it's happening right now – it's going to escalate sharply. It's always been evident that any effective measures to contain the threat of environmental catastrophe would have to be global in scope.

International efforts to avert catastrophe were advanced in the Paris negotiations with COP 21 in 2016. It should have gone into force in October 2016. The date was brought forward out of the concern that a Republican victory in 2016 might dismantle what had been achieved – not very much, but something. In fact, Republican denialism already had a significant impact. It was hoped that the Paris negotiations would lead to a verifiable treaty, but that hope was abandoned because the Republican congress would not accept any binding commitment.

What emerged was a voluntary agreement – evidently much weaker. In October 2016, a very significant agreement was reached to phase down the use of hydrofluorocarbons.[3] HFCs are super-polluting greenhouse gasses. It's delayed for a time for India and Pakistan, where increasing heat and awful poverty make cheap air conditioners which use HCFs a desperate necessity. The right response is evident. The rich countries should provide subsidies to accelerate the provision of non-HFC devices, the kind we use. Nothing like that seems to have been proposed, and if it were, its fate would probably have been the same as that of a verifiable treaty.

We might stop for a moment to ponder a most extraordinary fact: a major political organization in the most powerful country in the world's history is quite literally dedicated to the destruction of much of life on Earth. That might seem to be an unfair comment, but a little reflection will show that it's not. Right now, [in October 2016] we are reaching the end of the quadrennial electoral frenzy. In the Republican primaries, every single candidate denied the facts about climate change.

There was one exception, the "sensible moderate" John Kasich, who said, "Yes, it's happening, but we shouldn't do anything about it" – which is arguably even worse. So that's a 100% rejection. The winning candidate, as you know, calls for more use of fossil fuels, including coal, the most destructive. He is also for dismantling regulation, and for refusing funds to developing societies that are trying to move to sustainable energy such as, for example, non-polluting air conditioners in India – in every possible way, accelerating the race to disaster.

When you consider the stakes, it's a fair question whether there has ever been a more dangerous organization in human history than today's Republican Party. It's a fair question, and I think the answer is pretty clear. It's equally remarkable that these astonishing facts pass with virtually no comment. The campaign commentary about it proceeds at

A house burns during the Camp Fire in Paradise, California, November 2018.
Photo: Josh Edelson/AFP/Getty Images.

a level of vulgar triviality. We'll hear more of it in the presidential debates, with scarcely a glance at policy issues and almost nothing about the most critical questions that have ever arisen in human history – questions of literal survival *in the short term*. This is amazing blindness as the lemmings march to the precipice! In the past years, there has been extensive euphoric coverage of the prospects for energy independence – "a hundred years of energy independence" – with occasional

Floodwaters continue to inundate southwest Iowa. Floodwaters surround corn under a collapsed grain bin in this aerial photograph over Thurman, Iowa, on Saturday March 23, 2019. The deluge devastated much of the Midwest, making 2019 one of the worst years for flooding in the US.
Photo: Daniel Acker/Bloomberg via Getty Images.

comments on the local impact of fracking but scarcely a word pointing out that the euphoria amounts to an enthusiastic call for the sixth extinction to swallow us up as well.

Similarly, the growing threat of nuclear disaster, which is real and severe, barely elicits a comment. The two most important issues

August 31, 2005, a man pushes his bicycle through floodwaters near the Superdome in New Orleans after Hurricane Katrina left much of the city under water. Some of those who took shelter from Hurricane Harvey (2017) in Houston's Convention Center may have had flashbacks to this previous storm. Elected officials in Texas promised to heed the lessons from Katrina, which resulted in hundreds of deaths and tens of billions of dollars in damage.
Photo: AP Photo/Eric Gay, File

in all of human history, on which the fate of species depends, are virtually missing from the extensive commentary on the choice of leader for the most powerful country in world history and from the electoral extravaganza itself. It's not easy to find words to capture the enormity of this extraordinary blindness and perhaps words like these,

> … my imagination will not set bounds to the daring depravity of the times as the stock jobbers will become the pretorian band of the government, at once its tools and its tyrant; bribed by its largesses, and overawing it by clamors and combinations.[4]

Well, as you can tell from the quote's style, that's not from today, it's James Madison, in 1791, wondering about the fate of the new democratic experiment – not a bad description of the state it has reached 225 years later.

From the dawn of the nuclear age, there have been halting steps towards an international response that could contain the threat of nuclear war – or better maybe *terminate the threat* by eliminating these monstrous devices. One major step was the Non-Proliferation Treaty of 1968. In that treaty, the five nuclear states committed themselves to what were called good-faith efforts to eliminate nuclear weapons. Other signers pledged not to develop them. Three states with nuclear weapons have refused to sign, India, Pakistan, and Israel. All three have benefited in their nuclear weapons programs from US support: Pakistan during the Reagan years; India under Bush; Israel ever since a secret understanding which quickly became public between President Nixon and Israeli Prime Minister Golda Meir in 1969.

Well that's bad enough, but it could have been worse. In the 1970s, Henry Kissinger, Dick Cheney, Donald Rumsfeld, and other notables were urging US universities, primarily my own MIT, to aid Iran's nuclear programs. At the same time, high Iranian officials all the way up to the Shah were declaring quite openly that their goal was to develop nuclear weapons, but that didn't impede the efforts. Right after the Iran-Iraq War, the first George Bush as part of his program of cuddling his close friend Saddam Hussein went as far as inviting Iraqi nuclear engineers to the United States for advanced training in weapons production; that was in 1989.

All of this is better forgotten; we don't hear anything about it. Another international effort to contain the threat has been the establishment of nuclear-weapons-free zones. There is one in the Western Hemisphere, which excludes the United States and Canada but includes everyone else. There are others in Africa and in the Pacific. They are almost operative but not quite. They are still blocked by the US refusal to relinquish nuclear weapons in Diego Garcia and in the Pacific islands. The most important case by far would be in the Middle East. That's an initiative that has been led by the Arab states for over 20 years, but right now it's spearheaded by Iran. That would be the obvious way to eliminate any threat anyone believes might come from an Iranian nuclear weapons program. Strikingly, Iran is leading the efforts to institute a verifiable, nuclear-weapons-free zone.

The United States and Britain have a unique commitment to this initiative: when they were attempting to concoct some sort of pretext for invading Iraq, they appealed to a 1991 Security Council resolution which banned nuclear weapons in the Middle East. Ignored was the fact that the resolution explicitly commits the United States and Britain to work for a nuclear-weapons-free zone in the Middle East.

The efforts to carry this proposal forward have been regularly blocked by Washington, most recently by Obama in 2015 – transparently to prevent Israel's nuclear arsenal from coming under inspection. Apart from its significance in itself, the failure to move forward on a nuclear-weapons-free zone in the Middle East imperils the Non-Proliferation Treaty – the most important of all arms control treaties. It has been indefinitely extended, but that indefinite extension is conditioned on pledges to establish a zone free of weapons of mass destruction in the Middle East.

Protecting Israel's nuclear arsenal from inspection is evidently a high enough priority, so it justifies a threat to the major arms control treaty. Other facts unfortunately are not being discussed. The goal of abolishing nuclear weapons is not a utopian dream; it has in fact been forcefully advocated by quite mainstream establishment figures – among them Ronald Reagan's Secretary of State, George Shultz, former senator Sam Nunn, who was the Senate's leading specialist on nuclear weapons for many years, Henry Kissinger, and William Perry, one of the most respected analysts, along with having been a Secretary of Defense with long experience at what he calls "the nuclear brink."

These four were the *four* signers of an op-ed in the *Wall Street Journal*, which called for elimination – *total* elimination – of the scourge of nuclear weapons.[5] Another highly respected nuclear security expert, Bruce Blair, has formed a new organization called Global

Zero, which calls for an international treaty banning nuclear weapons. The International Court of Justice – the world court – came pretty close to this stand in a historic 1996 advisory opinion on the legality of possessing the threat or use of nuclear weapons.[6]

This month [October 2016], for the first time, the United Nations is considering a resolution to launch negotiations [towards] a legally binding instrument to – to get the words correctly – a *legally binding instrument to prohibit nuclear weapons leading towards their total elimination.* The resolution is sponsored by Austria, Brazil, Ireland, Mexico, Nigeria, South Africa. It's expected to gain the support of more than 120 states but without large-scale public support – and that's where your responsibility comes in – it will pass into oblivion just like other lost opportunities.[7]

The same holds for the steps that should be taken right now to reduce the international tensions that are escalating the threat of nuclear war to quite dangerous dimensions. This growing threat has elicited considerable alarm in the national security circles. William Perry has warned, in his words, that "we are facing nuclear dangers today that are in fact more likely to erupt into a nuclear conflict than during the Cold War." Perry is far from alone. Every year a group of experts organized by the atomic scientists updates the Doomsday Clock, established in 1947 at the dawn of the nuclear age, in which midnight means terminal disaster for everyone.[8] Two years ago [in 2014], they moved the clock three minutes closer to midnight, where it remains.[9]

That is the closest it's been since the early 1980s, when there was a very serious war scare. It's an event that should be better known and understood. At that time the Reagan administration launched an operation, "Operation Able Archer," if you want to look it up. It was designed to probe Russian defenses by simulating attacks, including nuclear attacks. This happened to be at a time of great international tensions. Right then, advanced missiles, Pershing 2 missiles, were being placed in Europe, in fact in Germany within a couple of minutes' – 10 minutes flight time to Russian territory. There were other rising tensions at the time. A couple of years ago, some Russian archives were declassified, and it was learned that the Russians took Able Archer very seriously. There has been uncertainty, however, about just what was understood in Washington. The CIA had claimed that the Russians didn't pay any attention; they knew it was just an exercise.

However, newly declassified documents just released have revealed that Washington understood right away that Able Archer was bringing the world to the verge of terminal war. These newly declassified documents reveal that US intelligence determined that the Russians were mobilizing forces, as they put it, into an unusual level of alert. According to protocol, that meant that the United States should have reacted in kind. One high-ranking US Air Force officer, Leonard Perroots, decided on his own *not* to follow prescribed procedure and do nothing – just to forget it – very likely averting a terminal nuclear war.

We already knew that, shortly after this, Russian automated systems detected an apparent massive US nuclear attack. The officer-in-charge, Stanislav Petrov, also decided to do nothing instead of transmitting the information to a higher level and possibly triggering a massive nuclear strike. These two gentlemen, Leonard Perroots and Stanislav Petrov, belong on the roll of people who've on their own blocked terminal nuclear war. They join Vasily Arkhipov, the Russian submarine commander, who in 1962, I think, in a dangerous moment of the Cuban missile crisis, decided again on his own to countermand an order from Russian submarines which were under attack. He decided to countermand an order to send off nuclear-tipped torpedoes, which again very likely would have escalated into terminal war.

It's on such decisions that the fate of civilization has relied all too frequently in the nuclear age, and it cannot go on. Today, the Doomsday Clock has been moved to three minutes to midnight just as during the time of Able Archer. The reasons given by the groups of experts were the growing threat of nuclear war and also, for the first time, the failure by government to deal in a serious way with the impending environmental crisis – the two major threats to survival that initiated the new era immediately after World War II.

The primary nuclear threat today is at the Russian border. Both sides are engaged in dangerous military build-ups, carrying out highly provocative acts and also expanding sharply their military arsenals. On the US side, one element is President Obama's proposal for a trillion dollar upgrading of the nuclear weapon systems that includes *new* nuclear weapons, cruise missiles, nuclear tips. These are recognized to be particularly dangerous because they can be scaled down to tactical battlefield use, which means that an officer on the ground would be tempted to use them, something that could quickly escalate to full-scale nuclear war.

Hillary Clinton, in a secret conversation that was leaked, raised questions about whether this should be continued. There, again, popular pressure can make a big difference. Also highly provocative, is an $800 billion missile defense system that Washington has recently installed in Romania allegedly in defense against non-existing Iranian missiles. It's recognized, in Russia, obviously, and it's recognized on all sides that what's called "missile defense" is basically a first-strike weapon.

It might conceivably impede a retaliatory strike. The Romanian installation is highly threatening to Russia; we'd of course never tolerate anything similar anywhere near our borders. The threat of war on the Russian border is in part, in large part, an outgrowth of NATO expansion since the collapse of the Soviet Union 25 years ago. That expansion ought to elicit more thought and discussion than it does. This was during the administration of the first president Bush and his Secretary of State, James Baker, and in Russia, of Mikhail Gorbachev.

Looking back at that time, the two sides had conflicting visions of the world order that should arise with the disappearance of the Soviet Union. Gorbachev called for a dismantling of all military alliances – of course the Warsaw Pact disappeared – to be replaced by a

Eurasian security system, integrating the former Soviet Union and Western Europe. That was Gorbachev's vision, but Bush and Baker had a different plan: NATO would expand while the Soviet system collapsed. And that's what happened.

The immediate issue was joined over the fate of Germany for obvious reasons. Gorbachev agreed to German unification – even German accession to NATO's hostile military alliance, which is a pretty remarkable concession in the light of recent history. Just in the past century Germany has practically destroyed Russia several times. There was, however, a *quid pro quo*, namely NATO would not expand "one inch to the east" was the phrase that was used – that meant East Germany.[10] Bush and Baker agreed to that compromise but only verbally.

It was a gentleman's agreement; it was not in writing. NATO immediately expanded to East Germany, but Bush and Baker correctly stated that they were not violating a *written* pledge, just a gentleman's agreement. Well, there is rich and interesting scholarly literature which has been trying to determine just what happened during that period. There were some crucial open questions like exactly what Bush and Baker had in mind. These questions have been pretty persuasively answered in an important way in a recent issue, a couple of months ago, of the MIT-Harvard journal *International Security*, by Joshua Itzkowitz Shifrinson.[11]

He did extensive new archival studies, and these revealed quite persuasively that the Bush-Baker verbal commitments to Gorbachev were quite explicitly designed to mislead Gorbachev while US dominance extended to the East. It's an important discovery. It shouldn't be hidden in a scholarly journal. Well that was only the first step. Under Clinton, NATO expanded further to the east, right to the Russian border. In 2008 and tentatively in 2013 under Obama, NATO even offered membership to Ukraine, which is in the Russian geopolitical heartland with long historical cultural relations with Russia, a highly provocative move.

George Kennan and other senior statesmen had warned earlier on, right at the beginning, that NATO enlargement was, as Kennan put it, "a tragic mistake" – a policy error of historic proportions, and we are now seeing the results. It's contributing to rising tensions on the Russian border that's the traditional invasion route through which Russia was virtually destroyed twice in the past century by Germany alone. The risk of terminal war is not slight. Reflecting on this matter, a European historian, Richard Sakwa, writes that NATO's mission today is to manage the risks created by its existence, which is in fact correct.[12]

Meanwhile, NATO's official mission has been extended well beyond the official mission to control the global energy system – the pipelines, sea lanes – and unofficially to serve as an intervention force under US command as we've seen. Well the fate of NATO shines a bright light on the real nature of the Cold War and its doctrinal basis. NATO, of course, had been presented as necessary to hold off the Russian hordes. We heard that for 50 years. Come 1991, no more Russian hordes, so what happens to NATO?

What happens provides no slight insight into what was the actual operative policy for the years before. It supports an observation by Harvard professor and government advisor Samuel Huntington. Ten years earlier, in 1981, in his words, "You may have to sell [intervention or other military action] in such a way as to create the misimpression that it is the Soviet Union that you are fighting. That is what the United States has done ever since the Truman Doctrine," in 1947.[13] As the clouds lifted in 1991 with the collapse of the Soviet Union, further evidence surfaced that bolsters this conclusion, but again it's kind of hidden from the public, though in fact it is in the public domain.

The Bush administration, first Bush administration, then in office of course immediately developed the new national security strategy in [its] defense budget and it was interesting reading. They said that the huge military system must remain in place and not to protect ourselves against the Russians but because of what they called the technological sophistication of third-world powers. If you are a well disciplined intellectual, you don't laugh when you read those words.

They also insisted that it would be necessary to retain what is called the "defense industrial base." That means the government-supported system of intervention in the economy through places like MIT and others which create the high-tech economy of the future. Quite interestingly, they referred also to the Middle East, where they said we must maintain intervention forces aimed at the Middle East. Then came this interesting phrase: in the Middle East the major problems that we faced "could not have been laid at the Kremlin's door" – contrary to many decades of lying. All of that passed, as usual, without comment.

As the Berlin Wall fell in 1989, Samuel Huntington, pursuing his earlier logic, warned that Gorbachev's public relations may be as much a threat to America's interests in Europe as were Brezhnev's tanks – and the threat of Gorbachev's peace offers was indeed overcome in the manner just reviewed, and we are now facing the consequences.

Well humans are now facing the most critical questions that have ever arisen in their history, questions that cannot be avoided or deferred if there is to be any hope of preserving, let alone enhancing, organized human life on Earth.

We surely cannot expect systems of organized power, state, or private systems to take appropriate actions to address these crises – not unless they are compelled to do so by constant, dedicated, popular mobilization and activism. A major task as always is education. I have given a couple of examples before, important ones I think, and there are plenty of others. These are efforts to develop public awareness and concern about the nature and the enormity of the problems we face and what their roots are, including in our own decisions.

There is a companion task, the usual one, to confront the issues themselves. That can take many forms. It can draw from – contribute to – the success of critical educational initiatives. Our own country is of course the most important case: for one reason because of its unique power and influence – and also for the simple reason that it's our own. It's here that what we do can be most influential. Popular activism can be highly influential. We've seen that over and over: activist engagement for 40 years has placed environmental concerns on the agenda of policy makers – not yet sufficiently but nevertheless in crucial and significant ways.

News clipping showing the Vermont delegation to the June 12, 1982 million person nuclear freeze demonstration in Manhattan.

Credit: David McCauley, Director of the American Friends Service Committee office in Vermont.

Segment of the June 12, 1982 march in Manhattan.
Source: WagingNonViolence.org.

The huge popular mobilization opposing nuclear weapons development in the early 1980s was a major factor in terminating the huge threats that arose at that time, actually paving the way for significant, again, if only partial steps towards reducing the enormous dangers that they pose. And there are many other illustrations of what can be achieved by dedicated efforts to educate, to organize, and to act. Such achievements are also based on the public's understanding of what has already been achieved and what actions can contribute to deepening and extending it.

Also, [there are] many illustrations of how the impact of popular movements can be magnified if they find ways to unify, to integrate their commitments. We share common goals of peace and justice. But there are challenges. These are, to be sure, difficult to overcome. There are powerful pressures that drive today's crucial issues to the margins of concern and discussion.

There are cultural and sociopolitical problems of considerable significance. It's important to bear in mind that although the United States has been the richest country in the world for a long time, way back into the 19th century, it has always been a kind of cultural backwater. That was true up until World War II.

Of course, that changed dramatically in the post-war world, but much of the population remains where it was, culturally traditional and pre-modern in many respects. For example, for 40% of the US population, these crucial issues of species survival are of little moment because Christ is returning to Earth within a couple of decades and then all will be settled, *that's 40% of the population.*

Two-thirds of Americans believe that global warming is happening. Far fewer think that it's caused by human activities. Only 40% are "aware," in the words of the polls, that most scientists think that global warming is happening and probably many fewer are aware that it's not *most* scientists but *an overwhelming consensus.* Unfortunately, if you look at the polls over the past 10 or 15 years (prior to 2016), awareness is not improving. On the rising threat of nuclear war, the reasons for it, and the consequences of resorting to nuclear weapons, the available information about public opinion is not at all encouraging.

Meanwhile, for the victims of the neoliberal assault on the population of the past generation, short-term problems of just getting by displace fundamental questions about the fate of their children and grandchildren. The tasks ahead are daunting, and they cannot be deferred.

Notes

1 Reves, Emery (1945, 2015) *The Anatomy of Peace.* Andesite Press.
2 Edgeworth, M. et al. (2016) Second Anthropocene Working Group Meeting (Conference Report). *The European Archaeologist* 47, http://nora.nerc.ac.uk/id/eprint/513430/1/Conference%20report_anthropocene_text%20(NORA).pdf.
3 United Nations Environment Program, "Frequently Asked Questions Relating to the Kigali Amendment to the Montreal Protocol," November 3, 2016, https://ec.europa.eu/clima/sites/clima/files/faq_kigali_amendment_en.pdf.
4 "To Thomas Jefferson from James Madison, 8 August 1791," Founders Online, National Archives, accessed April 11, 2019, https://founders.archives.gov/documents/Jefferson/01-22-02-0017. [Original source: *The Papers of Thomas Jefferson*, vol. 22, 6 August 1791–31 December 1791, ed. Charles T. Cullen. Princeton: Princeton University Press, 1986, pp. 17–18.]
5 Shultz, George P., William J. Perry, Henry A. Kissinger, and Sam Nunn (2010, January 19). "How to Protect Our Nuclear Deterrent." *The Wall Street Journal.* Retrieved from https://www.wsj.com/articles/SB10001424052748704152804574628344282735008.
6 International Court of Justice, "Legality of the Threat or Use of Nuclear Weapons." July 8, 1996, https://www.icj-cij.org/en/case/95.
7 United Nations General Assembly, Seventy-first session, "General and Complete Disarmament:

Taking Forward Multilateral Nuclear Disarmament Negotiations," October 14, 2016, http://reachingcriticalwill.org/images/documents/Disarmament-fora/1com/1com16/resolutions/L41.pdf. (Adopted October 27, 2016). On July 7, 2017, the United Nations General Assembly accepted the text of the Treaty on the Prohibition of Nuclear Weapons. The treaty has not yet entered into force. (Editors)

8 Bulletin of Atomic Scientists (nd) *The Doomsday Clock*. https://thebulletin.org/doomsday-clock/

9 In 2019, the Bulletin of Atomic Scientists set the Doomsday Clock to 2 minutes to midnight, where it remains at time of publication. See Chapter 5: The Third Threat.

10 Gorbachev's expectations were that NATO coverage would not extend into the territory of the former German Democratic Republic and beyond the borders of a unified Germany.

11 Shifrinson, Joshua R. Itzkowitz. "Deal or No Deal? The End of the Cold War and the U.S. Offer to Limit NATO Expansion." *International Security*, vol. 40. no. 4. (Spring 2016): 7–44. https://www.belfercenter.org/sites/default/files/files/publication/003-ISEC_a_00236-Shifrinson.pdf.

12 Sakwa, Richard (2016) *Frontline Ukraine: Crisis in the Borderlands*. London: I.B. Taurus.

13 Quoted in Chomsky, Noam (1992) *Deterring Democracy*. New York: Hill & Wang, p. 90.

2 Reaching People

SHAWN: *Great! You've laid out so much that it's hard to know where to begin. I think I will ask maybe a silly question. Most of the people in this room care about the future of the planet, of human beings. I think the first thing that we need to do if we are to preserve humanity is to convince the people who aren't in the room to care, but how do you do this? Is it that they don't know that these dangers are very real or is it that they don't have the ability to care that much about things that are intangible?*

NOAM: Just take a look at the examples that I mentioned and ask *yourselves*, not "the unwashed masses" out there. For example, there was a great fear for years about Iran, the so-called Iranian nuclear threat, but there is one very simple way to deal with it, very simple, institute a nuclear-weapons-free zone in the Middle East. Okay? Verifiable, nuclear-weapons-free zone as there are in other areas.

Will that be hard to convince Iran to agree to? No. They were leading the effort to institute it. Would it be hard to convince the Arab states in the region? No, they had been forcefully advocating it for over 20 years. In fact, they are the ones who insisted that unless this is done, they are going to pull out of the Non-Proliferation Treaty, destroying it. The United States and Britain, as I mentioned, had a unique commitment to this because of the resolution that they appealed to when they invented a pretext for invading Iraq.

How many people knew that? How many people know that this Iranian nuclear crisis could have been overcome, whatever the crisis was, very easily, no threats of war, no negotiations, no sanctions, just agree to institute a nuclear-weapons-free zone in the region. You can't do it just by clicking your fingers, but it's obvious how do it; it has been done elsewhere. Nobody knows, so you can't be committed to it. If people knew and they were committed to it, they might have compelled the United States' President Obama to withdraw the US opposition, which blocks it regularly.

If you read the professional arms-control journals, you know about this and you also know what's obvious on the face of it, that the United States is blocking this and indeed imperiling the Non-Proliferation

Treaty because it doesn't want Israel's nuclear weapons to be inspected. Is that interesting? Would people care about that when their lives depend on it? I suspect so. Take what is happening right in this moment as I mentioned, the United Nations for the first time is considering, right now in 2016, a resolution backed by major states – Brazil, Austria, others – that is probably going to get the votes of at least 120 countries calling for a treaty that declares these horrible devices illegal and demands their elimination.

Of course, the United States is not going to vote for it unless there is popular pressure to do so, but there can't be popular pressure to do so unless people at least know it's happening. This goes case by case. All through these matters – *the fate of NATO, what NATO was about, why it's expanding, why not accept Gorbachev's vision of a peaceful integrated Eurasia security with no military blocs* – people can't press that unless they know about them.

Take this leaked private conversation of Hillary Clinton and a number of funders recently. I think it came from WikiLeaks and was published in the *New York Times*. Clinton took some pretty reasonable positions like any politician adjusting to the audience to whom they are speaking – who happen to be powerful people but opposed to nuclear weapons. What she said was, "We should reconsider Obama's trillion dollar modernization program and we should oppose the development of the most dangerous parts of it." The ones I mentioned, *small* – they are called "small nuclear weapons" – and huge ones – the cruise missiles with nuclear warheads that have the dangerous capacity that you can scale them down to actual battlefield use. This means that an officer in the field whose forces are in danger could decide to use tactical nuclear weapons which could very quickly lead to terminal nuclear war. She said she's against it and we should reconsider the modernization program. When she talks to the next audience, she won't want to say that, but it could happen if people organized and mobilized to keep her feet to the fire. Say, "Okay, nice comments, stick to them." That can be done but only if you are aware of it and, secondly, if you are willing to become active and engaged, otherwise it's faded away like many other opportunities.

SHAWN: *When you speak of activism, first you are speaking of spreading awareness, why is it that the regular media, the* New York Times, *whatever, is so reluctant to write about these things in a way that would enable activism to reach more people. An activist seems like an idiot if he is saying things that no one has heard before.*

NOAM: There are a lot of reasons; it's not necessarily bad faith. If any of you went to journalism school, you know that one of the things that is taught and very highly respected is a concept called objectivity. "Objectivity" has a meaning; it means reporting accurately and fairly what's going on *inside* the Beltway, White House and Congress. You report that accurately and honestly, so if Donald Trump tweeted some obscenity at 3am, it's the lead story in the *New York Times,* and if Clinton said whatever she said, that's the big story.

If anything isn't being discussed within the narrow political economical establishment, you are not supposed to report it, that would be biased, that would be emotional, there are many terms for it, it's not being objective. That's a doctrine of journalism. If you think it's different in the academic world, you are not wrong. It's a *little* different. For example, the article in *International Security* that I mentioned, which is extremely significant, you look at the archival research: it's reported there in a rather dry way, but if you read it, what it says is, George Bush number one, the "judicious" Bush, and James Baker were purposely deceiving Gorbachev when they told him that NATO was not going to expand to East Germany. They were doing it in a way which would be purposely deceptive so that instead of dismantling NATO and accepting the vision of a world free of military blocs, we would be expanding NATO soon up to the Russian border with consequences that we see right now. It's there, but not there in a way which even the academic world sees them. Somebody might look at the article and say, "Kind of nice," and go on to the next article. You have to find the things that matter and bring them to public awareness. They are not hidden; it's a very free country, things are open. We find out a lot of things, but it's not handed to you on a silver platter, that's what activism is about.

SHAWN: Do you believe in talking to people who totally disagree with you or do you think it's best to speak to people who are slightly disagreeing with you but ready to easily listen?

NOAM: I don't think many people disagree about trying to save the world in which their grandchildren can survive. I think

New York – Prior to 2019, the largest mobilization against climate change involved hundreds of thousands of participants – 310,000 when demonstrators of all ages joined the "People's Climate March" of September 21, 2014.

Photo: Viviane Moos/Corbis via Getty Images.

people agree on that, so you talk to almost everyone, anyone who's moderately sane. What people do not know tends to lead them to highly irrational decisions – rational within their own framework, but the framework is missing crucial facts. I don't think it's quite possible to reach *every* audience, I don't expect to reach the Harvard Faculty Club, for example, but *most* people, you can reach them.

I think because they have shared interest, very simple shared interest like the simple fact that human beings have been around for 200,000 years and that the current generation has to decide whether that's going to continue. It's a pretty simple fact, a lot of evidence for it, overwhelming evidence. People, if they are brought to think about it, will care about it.

SHAWN: *What do you think of civil disobedience, chaining yourself to things and going to jail? Is that…?*

NOAM: Well I myself have been involved in it many times, been in and out of jail a lot of times, faced a long jail sentence, and so on. I think it's a legitimate tactic, but the ways in which it is conducted in my opinion often aren't legitimate. It's often conducted as kind of a statement of personal conscience. I'm going to take the risk because of my conscience – for some, my relation to God or something – whatever the consequences. I don't think that's the right way.

Civil disobedience makes sense if it brings to other people a recognition that there is something serious enough for some people to take risks and maybe they should think about it and go on and do something themselves. If the groundwork is laid, then civil disobedience can be an effective tool. If there is no groundwork, it's just not; it's harmful in fact. I should say this includes acts of people I very much respect and admire, close friends. For example, when peace activists break into a submarine base and bang missile nose cones without any preparation; the net effect is to anger the workers. "Why are you taking away our jobs?" The anger of other people, "Why are you getting in our way and annoying us?" What's the point? Just because it makes you feel good? That's not the right kind of civil disobedience. That has arisen many times, and if you have been involved in activism over the years, you know very well that these questions constantly arise, so it's pretty dramatic. For example, during the Vietnam War, I remember discussions with Vietnamese right at the peak of the war, in which they were talking about the kind of actions that they wanted to see, and they would give examples – like a group of women standing silently at the graves of American soldiers – that they really respected.

When you brought that to American activists, they just laughed. A lot of young people wanted to go down Main Street and break windows to show how much we hate the war, which of course just builds up support for the war. The Vietnamese wanted survival; they didn't care if that felt good. Those are questions that have to [be] asked constantly, *all the time.* You have to ask about the likely consequences of your act; whether or not it makes you feel good is not relevant – in fact, it is negative.

SHAWN: *Many of the people who do know about the consequences of nuclear war and climate change are quite well-educated people who are resented by a lot of people. Do you have any thoughts on how, I mean there is a class difference that Trump supporters who laugh at the idea of global warming and climate change have a built-in resentment toward people who've been well educated and who may be better off economically. How do we reach them?*

NOAM: That's serious. That is a very interesting phenomenon; it has to be dealt with sensitively and with understanding. As I mentioned, 40% of the population say it can't be a problem because of the Second Coming. Now that's a deep cultural problem in the United States. People who know something about US history should all... *we* should all understand it.

It's very important to realize that this country was a cultural backwater until World War II. [Until then,] if you wanted to study physics, you went to Germany. You wanted to become a writer, an artist, you went to Paris. There were exceptions of course but it was overwhelmingly true, and it was true even though the United States was far and away the richest, most powerful country in the world and had been for a long time. [There are] all kinds of historical reasons for that: it's a very insular country. There aren't many countries where you can travel 3,000 miles and be in about the same place where you left, not running into any different culture or language or anything like that. Protected by oceans, we keep away from those bad guys, enormous internal resources which nobody else had. There were a lot of waves of immigrants that became integrated and so on, so there are a lot of reasons for it, but it's there and you can't ignore it. You can't ignore it, and there is no point railing about atheism. These are issues that have to be understood, and it has to be understood that the churches really mean something to people, plenty of people, including the Trump supporters.

These are people who have just been cast aside, nobody does anything for them. The Democrats abandoned the working class decades ago. Republicans may take a populist line, but they are much more opposed to working people than even the Democrats in policies. Working-class males are, we are supposed to call them "middle class" in the United States, the phrase "working class" is a four-letter word here, but working-class males who are supporting Trump are actually supporting policies which are going to devastate them. Just take a look at the economic policies, the fiscal policies and others. But it's true that they are cast aside, and their values are being attacked. Their values are in many ways culturally traditional and pre-modern in the Western sense, but they are being attacked. One of the few refuges they have is the church. They are the church in a traditional community so you can't just laugh at it, it's serious. It has to be dealt with.

There is a very interesting book that just came out by Arlie Hochschild, a sociologist, who went to a pretty terribly impoverished area in Louisiana and lived there for six years and studied the people sympathetically.[1] This is deep Trump country, and her results are quite interesting. For example, these are people who are being devastated by chemical and other pollution from the petrochemical industry, but they are strongly opposed to the Environmental Protection Agency. When she asks why, they have reasons. They say, "Look what is the Environmental Protection Agency. It's some guy from the city with a Ph.D. who comes out here and tells me I can't fish but he doesn't go after the petrochemical industries. So, who wants them? I don't want them taking away my job and telling me what I can do and speaking to me with the cultivated accents meanwhile I'm under attack by all this stuff."

These attitudes are serious. They are significant. They deserve respect and not ridicule, and I think they can be addressed. For example, I think that say in the 1930s, I'm old enough to remember, in many ways, it was kind of like now; poverty was much greater. The depression was much worse than the current recession. In fact, it was a much poorer country than it is now

I was very hopeful. My own family, many of them were unemployed working class; most workers were unemployed, but they were hopeful.

They had a sense that things are going to get better. There were labor actions, the CIO's organizing, there were left political parties, the unions were providing real services: a couple of weeks in the country, educational groups, workers education, ways for people to get together – somehow we'll get out of all this. That's lacking. It has become a very atomized society. People are alone in it: used to be their TV sets, now it's their cell phones or iPhones or whatever. They're very atomized, isolated, makes them feel very vulnerable.

These are the kind of things that can be overcome by organization and activism. My own feeling [is] that the Trump supporters and the Sanders supporters could have been a unified bloc. Proper approaches to the problem take effort, sensitivity, and understanding of the kind for example that Hochschild showed with her sympathetic account of where these people are coming from and why. It's easy, say in a *New Yorker*, to have a cartoon about Trump and how ridiculous it is, but that's missing the point. Maybe it looks ridiculous, but he is reaching people for reasons and *we should be interested in the reasons.*

Actually, it's the same story, to turn to something else, with young Muslims in the West who are joining the jihad movements. It's not enough to scream at them; there are reasons. If you look at the circumstances in their lives, you can see the reasons and they can be addressed.

SHAWN: Real activism would start with a compassionate traveling into these unknown territories of our own land that Hochschild went into. What we can do in Boston and New York is less important than what we might be able to do if we went to Louisiana and moved there for a significant amount of time?

NOAM: I don't think we have to go many miles to find this. For example, a couple of years ago, I happened to be asked to give a talk at a high school in Boston, it's called English High School, and the reason is because nobody there speaks English [as a first language]. There are maybe a dozen languages with different immigrant groups. It's a very activist community. There are local activists who are discussing the kind of work that they do there, and it's important and interesting right here in Boston. People feel that it's hopeless, we can't do anything.

How can we fight these big powers? Some of the things that were described I think were very instructive for me; I think they would [be] for all of us. For example, get together a group of mothers who want to have a traffic light at a street where their kids have to cross when they go to school. They organize leaflets, talk to each other, talk to the local representatives and other things. Finally, they get their traffic light and that's empowering. That tells you that you can do something. We are not alone. We can do other things and then you go on from there; that's how things develop. Yeah, Louisiana, but not far from home either, and there's plenty [to do] incidentally – right in our own "educated" communities. The lack of understanding in educated circles is appalling. Almost everything I talked about tonight, for example, I doubt if a tiny fraction of well-educated academics who work in these areas would even know about it – that's right, *exactly* where we live.

Note

1 Hochschild, Arlie Russell (2016) *Strangers in Their Own Land: Anger and Mourning on the American Right.* New York: The New Press.

3 Thinking Strategically

RAY: My name is Ray Matsumiya. My family is from Hiroshima. My grandfather was [a] victim of the atomic bomb. That has great personal meaning to me. I run an organization that brings teachers from around the world to Hiroshima to understand the impact of the atomic bomb. The idea is that by spreading awareness to young people, these people can become future activists. They [may] stop the madness of nuclear weapons. My question to you is about location and priority. There are two streams of thought about this; the first is that it is very important to reach activist populations in countries that already have nuclear weapons. For example, in America, by mobilizing grassroots movements to question why a trillion dollars is being spent on modernizing nuclear weapons and creating obstacles to funding these expenditures. The other stream of thought is to work with countries that don't have nuclear weapons – like the resolution that's going on right now in the UN – with 120 countries that don't have nuclear weapons but they know that their future is tied to the use of nuclear weapons. Even if there is a limited nuclear exchange, it's going to impact the entire world and that's the rationale behind this stream. My question is, is it more important to activate populations in countries that have nuclear weapons or in ones that don't?

NOAM: I don't think it's either-or; they are all needed. Of the countries that have atomic bombs there are two categories; one consists of the signers of the Non-Proliferation Treaty. The official members – United States, Russia, China, France, and England – they have nuclear weapons. They are signers of the Non-Proliferation Treaty. They are bound by the Non-Proliferation Treaty Article 6 to carry out good faith efforts to eliminate the weapons and they should be doing it.

There are many ways to do it. As I mentioned in the talk, it's not a utopian hope. There are major, established figures who are calling for that, and we can join them. There are ways to progress towards it. You don't have to do it in one fell swoop. It has to be done mutually. It's an international activity. All of this has to be international; that's why the title is "Internationalism or Extinction." It can be done; for example, there has been some success since the end of the Cold War – even before – in reducing the number of nuclear weapons.

They are way in excess of any conceivable, imaginable, deterrents – so reducing them is significant. Eliminating the most dangerous ones is significant. For example, one thing that ought to be done is in the United States on what's called a "triad" of land-based, sea-based, and air-based missiles. The land-based ones are known by strategic analysts to be both useless and dangerous. They are useless because they are too slow, and you can't protect them and so on and so forth. They are dangerous because they are *targets*. They are of no use whatsoever. The sea-based ones can destroy the whole world a million times over. In fact, people who have investigated the [land-based] Minuteman bases say that the soldiers who are in charge know perfectly that this is a dead end. There is no point in these things. They don't pay any attention to it. They don't bother guarding them. They are off doing something else. It's a very dangerous phenomenon and useless. Okay, they can be closed, that could be a step towards inducing the Russians to drop some comparable thing.

Things like nuclear-weapons-free zones can be developed, and I think it could be done in the Middle East – if the US wasn't blocking it. The US is blocking it because people don't do anything about it. There are many other steps you can think of. Among the non-NPT states – well three of them have nuclear weapons, Pakistan, India, Israel – it's important to ensure that they get rid of their nuclear weapons or at least join the Non-Proliferation Treaty. Unfortunately, the United States has been supporting their nuclear weapons programs.

That's our problem. That's our fault. We are allowing that to happen, but we don't have to, okay? The other states are quite right to want these things eliminated; that's why there is a resolution in the UN right now from major states. It might bring in well over 100 states calling for their immediate banning, and that should be supported, not suppressed. It's not that it's secret. You can find out about it if you look at the UN records, but it's basically [a] secret that can be broken open and made public.

I think all of these things are possible and should be done in parallel; different people have different commitments, associations, interests, nobody can do everything, but you can pick what matters [to you] and do that.

EMILY: I work with 350 Mass. As someone who works on climate change, I think a lot about overcoming despair, and the picture you painted tonight is pretty bleak; it seems rightly so. But I'd love to hear your thoughts on how we sustain hope and commitment in the face of that. Are there historical parallels that you in particular think we can draw on as we face really unprecedented threats?

NOAM: It's pretty easy actually, [for me]: just think of the alternative. Suppose one does not maintain their spirits and one lets things happen – what's the world going to look like? There won't be any organized human life around, and most other species will be destroyed. Is that the world we want?

JASON: I'm Jason Pramas with the Boston Institute for Nonprofit Journalism. Professor Chomsky, understanding that the main push for banning nuclear weapons has to take place in the national and international stages, do you think it will be useful to organize for cities, states, and regional nuclear-weapons-free zones, including provisions that would ban nuclear weapons research and development at schools like MIT and related research institutes like Draper Lab?

NOAM: Yes, I think that's kind of analogous to nuclear-weapons-free zones, it can be done locally too. Banning nuclear-weapons research is very important. Incidentally at

MIT the story is a little different. This issue was brought up by activist students quite rightly during the period of major activism in the late '60s, and the outcome was kind of interesting. There were confrontations building up that finally led to the usual thing, a commission, a faculty-student commission. I was on it actually.

It investigated military work on campus; there was no nuclear-weapons work, but there was military work on campus, and it was pretty serious because the Institute was maybe 90% funded by the Pentagon. So, the natural question arose, and the results were interesting: on campus there was no classified work. Off campus there were two laboratories, Lincoln Labs and what's called Draper Labs. It was about 50% of the whole budget roughly, and they were doing military work off campus.

"Off campus" is a little bit of a formalism – nothing stops people from attending seminars across the street, but they were technically separate. Actually, the only military work on campus was not investigated incidentally because it was not in the science departments; it was in the *political science* department, which was actually doing what was effectively counter-insurgency research in Vietnam. They didn't call it that, it was the "Peace Research Institute" – *naturally.* It amounted to counter-insurgency studies in Vietnam, but that was about it.

There was a debate in the commission as to how to deal with it. It kind of sorted itself, roughly speaking, into conservatives, liberals, and radicals. The conservatives said they want to keep it the way it is. The liberals said, "we want to make a break, keep the military labs off campus." The few radicals agreed with the conservatives, "we should keep it on campus so it's a constant source of activism and education, not formally remove it where you can pretend it doesn't exist." But, the liberals won, so the research is technically off campus.

Similar questions can arise elsewhere. Actually, other questions have arisen too. Take Iranian nuclear weapons; it's a very interesting story. In the '70s when Iran was an ally, in fact the main US ally in the Middle East, "Guardian of the Gulf," it was called. There was strong pressure from Rumsfeld, Cheney, and Kissinger on MIT particularly to invite Iranian nuclear engineers to study nuclear engineering. We didn't know at the time what we should have known and know now: that the Iranian government [under the Shah] was openly declaring its intention to develop nuclear weapons.

It was hard to believe that Kissinger and those guys didn't know, but it wasn't public knowledge. There was turmoil on campus, which was pretty interesting. Students began to mobilize about it, and there was finally a student referendum where I think, my recollection is… I may have the numbers wrong so this is rough, about 80% of the students wanted to prevent it. It became a big enough issue so that there was a huge faculty meeting – people usually don't go to faculty meetings, they are too boring, but when there is a big issue, everybody comes.

The huge faculty meeting, during the big debate there weren't many – maybe five of us – who agreed with the students. Overwhelmingly, the faculty disagreed and said we should keep it on campus, which is kind of interesting if you think about it because the faculty are the students of 10 years earlier. That institutional change led to a dramatic change of attitude and not because of more information – faculty didn't have any more information than the students did. It's kind of an interesting example – people like us should think about it – of how our institutional roles influence the way we look at the world. Well, it went through, and [of] the people running the Iranian nuclear system, many were trained at MIT.

KIRKLAND: *Hello, I think two things are quite obvious and apparent from the discussion so far: one, we cannot rely on the opulent to fix these problems. Two, organizing ourselves into people's movements is the solution. I think those are two things that I've deduced from what we've talked about so far. My question is – once we do organize – what are the strategic pressure points? Have those been identified? I'm assuming that the strategy would be non-violence simply because of Stephan and Chenoweth's article in* International Security.[1] *I believe in the same journal you were referring to before, that non-violence is more effective, but if we do or when we do resist, when we do civil disobedience, what are the pressure points? Do we target the banks funding these things? Do we target local offices, the federal government?*

NOAM: I don't think there is a one-size-fits-all answer to this. It depends on who you are, what your circumstances are, what your interests are, who your associates are, what kind of talents you have, what kind of things you are good at, what kind of things you don't like to do. All of those pressure points matter – *all of them* – even down to a traffic light for your kids to walk across the street, where you learn you can organize. There are all kind of things that can be done, and it really is just an individual matter to find out what is the kind of thing that *I* think *I* can do that could be effective. We have to make our choices just like finding your own path in life; nobody can tell you how to do it.

LYN: *There is a lot of talk these days about lesser-evil voting and choosing the candidate who is most likely to get elected but inflict the least amount of harm. I'm wondering, isn't a vote for the lesser evil, whoever it may be, just a vote to perpetuate the two-party system that has led us down the road we are on today?*

NOAM: Well, maybe half the population abstains, are they changing the system? They are having no effect. There is a lot of confusion about lesser-evil voting; the point is really trivial. It's a point of logic. Simple logic. If you happen to be in a swing state, it's not clear where it's going to go; you have to make a decision about who you think is the worse candidate. Okay? Suppose you've decided you think Trump is the worse candidate, same if you think Clinton is the worse candidate, but let's say you suppose Trump is the worse candidate, you have a very simple choice.

Am I going to vote for Trump or am I going to vote against Trump? If you don't vote, you are voting for Trump, and if you vote for a third party, you are voting for Trump. That's arithmetic, and you can't argue with arithmetic. If you take one vote away from say Clinton, it's equivalent to adding one vote to Trump. One thing we can't argue about is arithmetic, that much is clear. What you can say is not clear is who is the worse candidate, that you can debate – quite frankly I don't think it's much of a debate – but you can decide. Those are two separate issues, and in all of this fear about lesser-evil voting, all the articles you read, you'll notice they are constantly confused if you separate them, it's all straightforward. You find who is the worse candidate, you decide whether you want to vote for them, and in a swing state voting say for a third party is voting for the worse evil, that's logic, you can't handle that. About not supporting the two-party system by abstaining? You are not changing the two-party system. You *could* be changing it first of all by working within it, like the Sanders movement to try to change and modify it. Today, the Democratic program is probably the most progressive in decades since the New Deal.

It won't be enacted unless there is pressure, but that's what political action is. The other possibility is *actually* developing a third party. Developing a third party doesn't mean just showing up every four years and running somebody for president, that doesn't do anything. What you have to do is constant activism starting at a local level, elect people to school boards, the town councils, state legislators, all in time, that way you can develop the basis for a third party. Now in our political system which we inherited from England, there is a barrier against third parties.

We inherited the first-past-the-post system. In countries with proportional representation there are many more opportunities to develop independent parties, so one option is to move towards a proportional-representation system. Another option is to use the mechanisms that are within the two-party system that allow for the development of independent parties like fusion candidates, like the Working Families Party in New York. You can vote for them and that helps the party, but the votes usually go to the Democrats, that's the way of working within the system. The possibilities are to develop a meaningful third party or work to change the existing parties – but just staying away from it doesn't help. Half the population does that anyway.

Actually, it's interesting, half the population, it's a very interesting study, revealing study, about the early Reagan years, early '80s, by Walter Dean Burnham. He is a very good scholar who studies electoral politics.[2] He did, as far as I know, the only study of the non-voters, asking what their socio-economic profiles were; it was a pretty interesting result. It turned out to be pretty similar to the people in Europe who vote for social democratic or labor-based parties. Here they just don't vote. It tells you something, something significant. Incidentally, that wouldn't be true today because the social democratic and labor-based parties in Europe have collapsed.

SHAWN: *Well, they've said that was the last question, the higher authorities, that we must come to an end tonight.*

NOAM: It's a beginning!

Notes

1 Stephan, Maria J. and Erica Chenoweth. "Why Civil Resistance Works: The Strategic Logic of Nonviolent Conflict." *International Security*, vol. 33. no. 1. (Summer 2008): 7–44.

2 Burnham, Walter Dean (1982) *The Current Crisis in American Politics*. Oxford and London: Oxford University Press.

4 Updated Reflections on Movements

EDITORS: *How do you think the looming threat of extinction should affect the Left – and its vision and strategy of activism? Does extinction call for a new framing of justice struggles? A new militancy?*

NOAM: The urgency of "looming extinction" cannot be overlooked. It should be a constant focus of programs of education, organization, and activism, and in the background of engagement in all other struggles. But it cannot displace these other concerns, in part because of the critical significance of many other struggles, and also in part because the existential issues cannot be addressed effectively unless there is general awareness and understanding of their urgency. Such awareness and understanding presupposes a much broader sensitivity towards the tribulations and injustices that plague the world – a deeper consciousness that can inspire activism and dedication, deeper insight into their roots and linkages. There's no point calling for militancy when the population is not ready for it, and that readiness has to be created by patient work. That may be frustrating when we consider the urgency of the existential threats, which is very real. But frustrating or not, these preliminary stages cannot be skipped.

EDITORS: *Do you think movements should specifically focus on extinction, as the Extinction Rebellion in the UK is doing? What might that look like in the US? How should existing movements change their focus on strategy?*

NOAM: The UK Extinction Rebellion movement has laudable goals. In the US, the grassroots Earth Strike movement (earth-strike.com) is planning actions through 2019 leading to a "General Strike to Save the Planet" in September. Other organizations are developing plans as well. These are all very valuable initiatives, which merit strong support. Inevitably, their success will depend on the general raising of consciousness. We cannot ignore the realities of the world we are living in – a world in which, for example, half of Republicans, according to recent polls, deny that global warming is even happening, and of the rest, a bare majority consider that humans have some responsibility for it. According to the most recent general polls (March 2018), a mere 25% of Republicans

Extinction Rebellion, UK.
Photo: Getty Images.

"think global warming should be a high or very high priority for the president and Congress." A world in which regular reports on expansion of fossil fuel production in the liberal press hail our taking the lead with its implications for global power, perhaps mentioning some local environmental effects of opening new areas for exploitation (water shortages for ranchers, for example), but with scarcely a word, if any, on what this entails for the lives of the next generation. Same with regard to the second major threat to survival. There has been little comment about the [Trump] administration's new National Security Strategy, calling for trillions of dollars to ensure US "overmatch" – preponderance over any coalition of rivals – and assurance that the US can win a war against China and/or Russia, though a war with either would obliterate everything.[1] This was enthusiastically presented by the fabled "adult in the room," the "voice of reason" among the Trumpians – "Mad Dog" Mattis – along with plans for highly destabilizing new weapons and reversal of the slow progress towards mitigating the grave nuclear threat.

There's simply no way to avoid the steady hard work of developing consciousness and understanding. Innovative and dramatic actions can stimulate such awareness and recognition of urgency of action if integrated into broader initiatives. General outlines are easy enough to sketch; filling in the details with specific programs and actions is what counts.

EDITORS: *Where do you see the strongest hope for international solidarity and agreements coming from? Are there structural forces such as a shift in the nature of globalization? Is there hope in specific nations or institutions? Or in globalized social movements? How can they overcome the hyper-nationalism of the new regime of autocrats from Brazil to the Philippines to the US?*

NOAM: There is no simple formula. To confront the growing and destructive autocratic and hyper-nationalist tendencies we have to first understand their roots. The topic is too broad to address seriously here, but there is good evidence, I think, that a substantial factor is the neoliberal austerity programs of the past generation. These have concentrated wealth and undermined functioning democracy, casting much of the population aside, leading to understandable resentment and anger, often taking pathological forms, and leaving people prey to demagogues. These developments can only be countered by progressive social movements that offer credible answers to the often bitter exigencies of daily life and, even better, point the way to needed social and institutional change. That should be the basis for international solidarity, particularly in a globalized world where many face similar threats to decent existence and have opportunities for communication and interaction. These have been exploited effectively by international capital, but much less so by the victims of harsh policies. These [are] severe problems that have to be confronted and overcome.

EDITORS: You don't write much about capitalism in this book. Do you see capitalism driving extinction? Do we need to move beyond capitalism to ensure survival?

NOAM: The varieties of state capitalism that now exist are based on principles that I think should not be tolerated. Some of their dominant properties such as ignoring externalities and the drive for growth and damn the consequences virtually guarantee disaster. But in fact the systems are pliable enough to offer hope for survival through the development of green economies; for example, say, along the lines that economist Robert Pollin has spelled out in some detail.[2] That's fortunate because in the real world conditions are not ripe for large-scale institutional change to true democratization and popular participatory control of social, economic, and political life – though in fact seeds of such developments *do* exist and can flourish. Like it or not, the urgent issues of today will have to be confronted within the general framework of existing institutions, while at the same time serious efforts should be undertaken to rid ourselves of oppressive institutions and move forward to greater freedom, justice, authentic democracy, cooperation and mutual aid in all spheres of life.

EDITORS: *Following up, you have focused on the leading threats to human existence: fossil fuel-driven climate change and the potential for nuclear weapons conflict. This seems reasonable; however, you also allude to many other sources of species extinction. As we survey the scene, what stands out is how comprehensive the threats are and how corrective changes implicate every dimension of human life. How are we to understand the system that produces such an all-embracing character? Is there a system? Or a multiplicity of systems? To make this more concrete, how does someone who cares about, say, opiate abuse in the Rustbelt make sense of the challenges confronting their communities?*

NOAM: The world is a complex place, but we can identify systematic factors and structures. Take opiate abuse. Why is there opiate abuse in the Rustbelt? Why is life expectancy continuing to decline in the US for the first time since World War I and the flu pandemic? Why is this particularly salient among working-class whites – who have been cast aside by the neoliberal policies of the past generation, including the particular form of globalization that has been designed in the interests of the investor class and transnational capital? The move from opiate abuse to the regressive policies that began to take shape in the late '70s, accelerated by Reagan and his successors, is fairly straightforward. And as I mentioned earlier, there are counterparts elsewhere, in the rise of "illiberal democracy" and the collapse of the centrist forces that have dominated political life since World War II. Studies ranging from the US to Sweden and others have found that xenophobia, anti-immigrant hysteria, racism, and the rise of the ultranationalist right in general have tended to follow economic distress and reversal of social-democratic programs. So, yes, the world is a complex place and a multitude of factors interact, but there are some systematic features in global malaise – which point the way to corrective action.

EDITORS: *There are many proposals that are gaining currency; perhaps the foremost among them is the Green New Deal. What kind of resistance should its proponents expect from corporate power, and the media and politicians in their employ? What organizational decisions should progressive activists make in terms of their own coalition building to overcome such resistance?*

NOAM: The main point to keep in mind is that proposals of this nature must succeed. *Must*, or else we are doomed. Some of the proposals are quite carefully designed, and developed in a form that can be used as a basis for organizing, notably Pollin's work on a Green New Deal. There is of course ample reason to expect corporate resistance, both from the evidence of history and the nature of state-capitalist markets. But it seems that we have passed beyond the days when, for example, ExxonMobil executives reacted to James Hansen's publicizing the threat of global warming in 1988 by devoting resources to engendering skepticism or outright denialism – knowing exactly what they were doing, since their own scientists had long been among the leaders in demonstrating the extreme gravity of the threats.

By now the threats are so evident that we seem to have moved to an era marked more by co-optation and mitigation rather than by outright rejection of reality. That's often happened in the past: the practices of the lethal tobacco industries, for example. The shift offers opportunities for activists, but they are littered with pitfalls that have to be recognized and avoided. Strategies should be designed to grasp the opportunities – to fail to do so is counterproductive – but with due attention to the motives, intentions, and manipulations of the systems of power. That's more difficult than confronting simple denialism but also provides openings for education and organizing, which must be intensified. There's no time to waste.

EDITORS: *Do you feel the extinction crisis should change anything about our social movements? For example, should they coordinate to make their work more effective in heading off extinction? Do they need to become more globalized to help achieve far-reaching climate and arms agreements? Do they need to highlight more the connection between corporate power and the threat of extinction?*

NOAM: How about all of the above?

For activists, there are strong temptations – understandable, valid – to devote intense efforts to critical issues in the immediate focus of their work. But linkages with other social struggles are real, not just at home but globally. All can gain by considered and careful initiatives to pursue "intersectionality" and solidarity. And all can gain by exploring and confronting the common institutional roots that in significant respects lie behind the problems, often crises, that are in the forefront of particular commitments. Capital is coordinated and globalized. The struggles against injustice and oppression must develop interactions and mutual support in their own ways. The dreams of a true International should not fade. And they gain overwhelming impact when we recognize the severe threats to organized social life that cast their grim shadow over all other concerns.

Notes

1 Department of Defense (2018) *Summary of the 2018 National Defense Strategy of the United States of America.* https://dod.defense.gov/Portals/1/Documents/pubs/2018-National-Defense-Strategy-Summary.pdf.

2 Pollin, Robert, Heidi Garrett-Peltier, Jeannette Wicks-Lim, Shouvik Chakraborty, and Tyler Hansen (2017) *Green Growth Programs for U.S. States.* https://www.peri.umass.edu/publication/item/1032-green-new-deal-for-u-s-states.

5 The Third Threat

The Hollowing Out of Democracy

On April 11, 2019, some 30 months after his 2016 pre-election talk, Noam Chomsky returned to the Old South Church to again address a capacity crowd on the theme of Internationalism or Extinction.[1] *Beginning with a personal reflection, he augments his description of the existential threats facing humanity to include the political process itself: the subversion of democracy by fossil fuel, corporate, and nationalist interests. – Editors*

If you'll indulge me, I'd like to start with a brief reminiscence of a period which is eerily similar to today in many unpleasant respects. I'm thinking of exactly 80 years ago, almost to the day; it happened to be the moment of the first article that I remember having written on political issues. Easy to date: it was right after the fall of Barcelona in February 1939.

The article was about what seemed to be the inexorable spread of fascism over the world. In 1938, Austria had been annexed by Nazi Germany. A few months later, Czechoslovakia was betrayed, placed in the hands of the Nazis at the Munich Conference.

In Spain, one city after another was falling to Franco's forces. February 1939, Barcelona fell. That was the end of the Spanish Republic. The remarkable popular revolution, anarchist revolution, of 1936, '37, '38, had already been crushed by force. It looked as if fascism was going to spread without end.

It's not exactly what's happening today, but, if we can borrow Mark Twain's famous phrase, "History doesn't repeat but sometimes rhymes" – too many similarities to overlook.

When Barcelona fell, there was a huge flood of refugees from Spain. Most went to Mexico, about 40,000. Some went to New York City, established anarchist offices in Union Square, secondhand bookstores down 4th Avenue. That's where I got my early political education, roaming around that area. That's 80 years ago. Now it's today.

We didn't know at the time, but the US government was also beginning to think about how the spread of fascism might be virtually unstoppable. They didn't view it with the same alarm that I did as a 10-year-old. We now know that the attitude of the State Department was rather mixed regarding what the significance of the Nazi movement was. Actually, there was a consul in Berlin, US consul in Berlin, who was sending back pretty mixed comments about the Nazis, suggesting maybe they're not as bad as everyone says. He stayed there until Pearl Harbor Day, when he was withdrawn – the famous diplomat named George Kennan. Not a bad indication of the mixed attitude towards these developments.

It turns out, couldn't have known it at the time, but shortly after this, 1939, the State Department and the Council on Foreign Relations began to carry out planning about the postwar world, what the postwar world would look like. And in the early years, right about that time, next few years, they assumed that the postwar world would be divided between a German-controlled world, Nazi-controlled world, most of Eurasia, and a US-controlled world, which would include the Western Hemisphere, the former British Empire – which the US would take over – parts of the Far East. And that would be the shape of the postwar world. Those views, we now know, were maintained until the Russians turned the tide. Stalingrad, 1942–1943, the huge tank battle at Kursk, a little later, made it pretty clear that the Russians would defeat the Nazis. The planning changed. Picture of the postwar world changed, went on to what we've seen for the last period since that time. Well, that was 80 years ago.

Today we are not facing the rise of anything like Nazism, but we are facing the spread of what's sometimes called the ultranationalist, reactionary international, trumpeted openly by its advocates, including Steve Bannon, the impresario of the movement. [He] just had a victory yesterday: the Netanyahu election in Israel solidified the reactionary alliance that's being established, all of this under the US aegis, run by the triumvirate, the Trump-Pompeo-Bolton triumvirate. I could borrow a phrase from George W. Bush to describe them, but, out of politeness, I won't. The Middle East alliance consists of the extreme reactionary states of the region – Saudi Arabia, United Arab Emirates, Egypt under the most brutal dictatorship of its history, Israel right at the center of it – confronting Iran. [There are] severe threats that we're facing in Latin America. The election of Jair Bolsonaro in Brazil put in power the most extreme, most outrageous of the right-wing ultranationalists who are now plaguing the hemisphere. Yesterday, Lenín Moreno of Ecuador took a strong step towards joining the far-right alliance by expelling Julian Assange from the embassy. He's picked up quickly by the UK, [and] will face a very dangerous future unless there's a significant popular protest. Mexico is one of the rare exceptions in Latin America to these developments. This has happened – in Western Europe, the right-wing parties are growing, some of them very frightening in character.

There is a counter-development. Yanis Varoufakis, the former finance minister of Greece, a very significant, important individual, along with Bernie Sanders, has urged the formation of a Progressive International to counter the right-wing international that's developing. At the level of states, the balance looks overwhelmingly in the wrong direction. But states aren't the only entities. At the level of people, it's quite different. And that could make the difference. That means a need to protect the functioning democracies, to enhance them, to make use of the opportunities they provide, for the kinds of activism that have led to significant progress in the past [and that] could save us in the future.

I want to make a couple of remarks below about the severe difficulty of maintaining and instituting democracy, the powerful forces that have always opposed it, the achievements of somehow salvaging and enhancing it, and the significance of that for the future. But first, a couple of words about the challenges that we face, which you heard enough about already and you all know about. I don't have to go into them in detail. To describe these challenges as "extremely severe" would be an error. The phrase does not capture the enormity of the kinds of challenges that lie ahead. And any serious discussion of the future of humanity must begin by recognizing a critical fact, that the human species is now facing a question that has never before arisen in human history, [a] question that has to be answered quickly: Will human society survive for long?

Well, as you all know, for 70 years we've been living under the shadow of nuclear war. Those who have looked at the record can only be amazed that we've survived this far. Time after time it's come extremely close to terminal disaster, even minutes away. It's kind of a miracle that we've survived. Miracles don't go on forever. This has to be terminated, and quickly. The recent Nuclear Posture Review of the Trump administration dramatically increases the threat of conflagration, which would in fact be terminal for the species.[2] We may remember that this Nuclear Posture Review was sponsored by Jim Mattis, who was regarded as too civilized to be retained in the administration – gives you a sense of what can be tolerated in the Trump-Pompeo-Bolton world.

Well, there were three major arms treaties: the ABM Treaty, Anti-Ballistic Missile Treaty; the INF Treaty, Intermediate Nuclear Forces; the New START treaty.

The US pulled out of the ABM Treaty in 2002. And anyone who believes that anti-ballistic missiles are defensive weapons is deluded about the nature of these systems.

The US has just pulled out of the INF Treaty, established by Gorbachev and Reagan in 1987, which sharply reduced the threat of war in Europe, which would very quickly spread. The background of that signing of that treaty was the demonstrations that you just saw depicted on the film [see http://ChomskySpeaks.org]. Massive public demonstrations were the background for leading to a treaty that made a very significant difference. It's worth remembering that and many other cases where significant popular activism has made a huge difference. The lessons are too obvious to enumerate. Well, the Trump administration has just withdrawn from the INF Treaty; the Russians withdrew right afterwards.

If you take a close look, you find that each side has a kind of a credible case saying that the opponent has not lived up to the treaty. For those who want a picture of how the Russians might look at it, the *Bulletin of Atomic Scientists*, the major journal on arms control issues, had a lead article a couple weeks ago by Theodore Postol pointing out how dangerous the US installations of anti-ballistic missiles on the Russian border – how dangerous they are and can be perceived to be by the Russians.[3] Notice, *on the Russian border.* Tensions are mounting. Both sides are carrying out provocative actions. In a rational world, what would happen would be negotiations between the two sides, with independent experts to evaluate the charges that each is making against the other, to lead to a resolution of these charges, to restore the treaty. That's in a rational world. But it's unfortunately not the world we're living in. No efforts at all have been made in this direction. And they won't be, unless there is significant pressure.

Well, that leaves the New START treaty. The New START treaty has already been designated by the figure in charge (who has modestly described himself as the greatest president in American history) the worst treaty that ever happened in human history, the usual designation for anything that was done by his predecessors. Trump added that we've got to get rid of it. If in fact this comes up for renewal right after the next election, a lot is at stake. A lot is at stake in whether that treaty will be renewed. It has succeeded in very significantly reducing the number of nuclear weapons, to a level way above what they ought to be, but way below what they were before. And it could go on.

Meanwhile, global warming proceeds on its inexorable course. During this millennium, every single year, with one exception, has been hotter than the last one. There are recent scientific papers, [by] James Hansen and others, which indicate that the pace of global warming, which has been increasing since about 1980, may be sharply escalating and may be moving from linear growth to exponential growth, which means doubling every couple of decades. We're already approaching the conditions of 125,000 years ago, when the sea level was about roughly 25 feet higher than it is today. With the melting, the rapid melting, of the Antarctic's huge ice fields that point might be reached. The consequences of that are almost unimaginable. I mean, I won't even try to depict them, but you can figure out quickly what that means.

While this is going on, you regularly read in the press euphoric accounts of how the United States is advancing in fossil fuel production. It's now surpassed Saudi Arabia. We're in the lead of fossil fuel production. The big banks, JPMorgan Chase and others, are pouring money into new investments in fossil fuels, including the most dangerous, like Canadian tar sands. And this is all presented with great euphoria, excitement. We're now reaching "energy independence." We can control the world, determine the use of fossil fuels in the world.

Barely a word on what the meaning of this is, which is quite obvious. It's not that the reporters, commentators don't know about it, that the CEOs of the banks don't know about it. *Of course they do.* But these are kind of institutional pressures that just are extremely hard to extricate themselves from. Try to put yourself in the position of, say, the CEO of JPMorgan Chase, the biggest bank, which is spending large sums in investment in fossil fuels. He certainly knows everything that you all know about global warming. It's no secret. But what are the choices? Basically, he has two choices. One choice is to do exactly what he's doing. The other choice is to resign and be replaced by somebody else who will do exactly what he's doing. It's not an individual problem. It's an institutional problem, which can be met, but only under tremendous public pressure.

And we've recently seen, very dramatically, how it can – how the solution can be reached. A group of young people, the Sunrise Movement, organized, got to the point of sitting-in in congressional offices, [and] aroused some interest from the new progressive figures who were able to make it to Congress. Under a lot of popular pressure, Congresswoman Alexandria Ocasio-Cortez, joined by Senator Ed Markey, actually placed the Green New Deal

Extinction Rebellion march arrives at Parliament Square, London on April 23, 2019. The demonstrators, after over a thousand arrests, targeted Parliament, as MPs were sitting this day after Easter recess. Photo: Alberto Pezzali/NurPhoto via Getty Images.

on the agenda. That's a remarkable achievement. Of course, it gets hostile attacks from everywhere: it doesn't matter. A couple of years ago, it was unimaginable that it would be discussed. As the result of the activism of this group of young people, it's now right in the center of the agenda. It's got to be implemented in one form or another. It's essential for survival, maybe not in exactly that form, but some modification of it. Tremendous change achieved by the commitment of a small group of young people. That tells you the kind of thing that can be done.

Meanwhile, the Doomsday Clock of the Bulletin of Atomic Scientists last January was set at two minutes to midnight. That's as close as it's been to terminal disaster since 1947. The announcement of the settlement – of the setting – mentioned the two major familiar threats: the threat of nuclear war, which is increasing, [and the] threat of global warming, which is increasing further. And it added a third for the first time: the undermining of democracy.[4] That's the third threat, along with global warming and nuclear war. And that was quite appropriate, because functioning democracy offers the only hope of overcoming these threats. They are not going to be dealt with by major institutions, state or private, acting without massive public pressure, which means that the means of democratic functioning have to be kept alive, used the way the Sunrise Movement did it, the way the great mass demonstration in the early '80s did it, and the way we continue today.

Notes

1 Video and complete audio from this event sponsored by the Campaign for Peace, Disarmament, and Common Security, encuentro5, Massachusetts Peace Action, and the Wallace Action Fund is available from the website http://ChomskySpeaks.org.
2 Office of the Secretary of Defense (2018) *Nuclear Posture Review.* https://media.defense.gov/2018/Feb/02/2001872886/-1/-1/1/2018-NUCLEAR-POSTURE-REVIEW-FINAL-REPORT.PDF
3 Postol, Theodore A. (2019) "Russia May Have Violated the INF Treaty. Here's How the United States Appears to Have Done the Same." *Bulletin of Atomic Scientists*, February 14. https://thebulletin.org/2019/02/russia-may-have-violated-the-inf-treaty-heres-how-the-united-states-appears-to-have-done-the-same/.
4 In the words of the *Bulletin*, "These major threats – nuclear weapons and climate change – were exacerbated this past year by the increased use of information warfare to undermine democracy around the world, amplifying risk from these and other threats and putting the future of civilization in extraordinary danger. There is nothing normal about the complex and frightening reality just described." https://thebulletin.org/doomsday-clock/

6 To Learn More

Documenting a lifetime as a public intellectual, Noam Chomsky's works are a unique repository for contemporary analysis and activism. Further exploration is warranted for many reasons, not the least of which is the multifaceted character of Noam's commitments. In his writings, one finds not only historical depth and analytical clarity but also a wide lens that pans across society's many sites of contention and reveals the many connections between them and also the sometimes obscured underlying political and economic structures. Two well-organized and searchable online starting points are:

- The Massachusetts Institute of Technology's site, unBox, the Chomsky Archive, https://libraries.mit.edu/chomsky/
- The Official Noam Chomsky website founded by Pablo Stafforini and organized by Valeria Chomsky, http://chomsky.info

An additional and original entry point for delving into Noam Chomsky's thinking examines *his* sources: the many works that

Noam has referenced and shared with his audiences as exemplary efforts. The Chomsky List, http://ChomskyList.com, adopts this approach and remains a valuable resource.

It also adds value by providing a *short* list of what may one day be termed "the Chomsky classics of social and political theory"; these include *Manufacturing Consent: The Political Economy of the Mass Media* (co-authored with Edward S. Herman, 1988) on the institutional structure of media and propaganda; *American Power and the New Mandarins: Historical and Political Essays* (1969), Noam's first "political" book that establishes his analysis of the US relationship with the rest of the world; *For Reasons of State* (1973) picks up on these themes and is Noam's definitive work on the Vietnam War; *Government in the Future* (1970), in which Noam presents his vision of a libertarian socialist future and its contrasts with statist forms of socialism and capitalism; *Hegemony or Survival?: America's Quest for Global Dominance* (2003) explores many of the themes presented in the present book and ties future prospects for humanity to how it contends with a ruling class that is militarizing every dimension of human existence; *Fateful Triangle: The United States, Israel and the Palestinians* (1999) is the definitive work establishing Noam's critique of the US-Israel relationship and the grounds for his solidarity with the Palestinian freedom struggle.

Of course, many books have also been written about Noam and his thinking. Among the best biographies, one that integrates Noam's scientific work on linguistics with his social and political thinking is James McGilvray's *Chomsky: Language, Mind and Politics* (2nd Edition, 2013). McGilvray has also edited an essential survey of Noam's thinking, *The Cambridge Companion to Chomsky* (2nd Edition, 2017). Similarly, Milan Rai's *Chomsky's Politics* (1995) provides an overview of Noam's political thought and its reception on the American intellectual scene. Robert Barsky's *The Chomsky Effect: A Radical Works Beyond the Ivory Tower* (2007) situates Noam's political work in its political context, thereby uncovering Noam's unique blend of argument and activist provocation.

Several Chomsky documentaries have gained a deservedly broad circulation. Best known of these is filmmakers Mark Achbar and Peter Wintonick's *Manufacturing Consent* (1992) and director Peter Hutchison's *Requiem for the American Dream* (2015). The companion video to this book (see www.chomskyspeaks.org/) overlaps with Chapter 1 of this text and is very suggestive of the atmosphere of the typical "Chomsky event."

Index

Note: *Italic* page numbers refer to figures and page numbers followed by "n" denote endnotes.

ABM (Anti Ballistic Missile) Treaty 91–2
Achbar, Mark 101
activism: and journalism 49–50; links between struggles in 84–5; popular 10–11, 39–40, 42, 70–1; *see also* social movements
Anthropocene 7–8, 15–17
Arkhipov, Vasily 10, 32
Assange, Julian 89
Austria 30, 48, 86
autocracy 79

Baker, James 34–6, 51
Bangladesh 19
Bannon, Steve 89
Barsky, Robert 101
Blair, Bruce 29
Bolsonaro, Jair 89
Brazil 30, 48, 79, 89
Britain 28, 47, 88
Burnham, Walter Dean 73–4
Bush, George H. W. 27, 34–6, 38, 51
Bush, George W. 89

capitalism 80, 100
Capitalistocene 8–9
carbon emissions 7, 13n1, 17
Cheney, Dick 27, 68
Chenoweth, Erica 70
China 63, 78
Chomsky, Noam, more about 99–101
Chomsky event 2–5, 101; organizations piggy-backing on 11–12
Churchill, Winston 15

civil disobedience 54–5, 70
climate change: and class resentment 55, 58; co-optation and mitigation of 83–4; despair about 65–6; mobilization against *52*, *96*; threat of 6–7, 81, 93–4, 97
Clinton, Bill 36
Clinton, Hillary 34, 48–50, 71–2
Cold War 12, 30, 37, 45, 64
consciousness, developing 78–9
Cuban missile crisis 9, 32

Darfur 19
defense industrial base 38
democracy: authentic 81; illiberal 82; protecting 90; undermining 6, 86, 97, 98n4
Democratic Party 57, 72–3
Diego Garcia 28
Doomsday Clock 31, 33, 45n9, 97
Draper Laboratory 11, 66–7
droughts 19

Earth Strike movement 76
education, critical initiatives in 39
Egypt 89
Einstein, Albert 15
electoral systems 73
elites, rational or enlightened 10
energy independence 24, 94
energy system, global 37
England *see* Britain
English High School 60
Environmental Protection Agency 57–8
externalities, ignoring 80
extinction, threats of 1, 6, 14–17, 75, 81, 84
Extinction Rebellion 76–7, 96

fascism 86–7
flooding *24*, *25*
forest fires *18*, *23*

fossil fuels 7, 22, 78, 86, 94–5
France 63

Germany: Nazi regime in 86–8; unification of 35, 45n10; US missiles in 31
global warming *see* climate change
Global Zero 29–30
globalization 79–80, 82
Gorbachev, Mikhail 34–6, 38–9, 45n10, 48, 51, 92
Great Acceleration 7–8
green economies 80
Green New Deal 83, 95–7

Hansen, James 83, 94
HFCs (hydrofluorocarbons) 21–2
Hiroshima nuclear attack 14, 62
Hochschild, Arlie 58, 59
Huntington, Samuel 37–9
hurricanes *20*, *25*
Hutchison, Peter 101
hyper nationalism *see* ultranationalism

ice fields and glaciers, melting 17–19, 94
India 21, 23, 27, 65
INF (Intermediate Nuclear Forces) Treaty 91–2
information warfare 98n4
international cooperation 9, 15
International Court of Justice 30
intersectionality 85
Iran 27–8, 34, 46–7, 68–9, 89
Iraq 27–8
Israel: nuclear weapons of 27–9, 47, 65; and reactionary alliance 89

journalism, objectivity in 50

Kasich, John 22
Kennan, George 36, 88
Kissinger, Henry 27, 29, 68

lesser-evil voting 71–2
Lincoln Labs 67
Louisiana 57–60

Madison, James 8–9, 26
Markey, Ed 95

Matsumiya, Ray 62
Mattis, Jim "Mad Dog" 78, 91
McCauley, David 40
McGilvray, James 101
Meir, Golda 27
Mexico 30, 87, 89
Middle East: nuclear-weapons-free zone in 28–9, 46–7, 65; US intervention forces in 38
militancy 1, 75–6
missiles 10, 31, 33–4, 49, 64, 92
MIT, Pentagon-funded research at 11, 67–9
Moore, Jason 8
Moreno, Lenín 89

NATO 34–7, 45n10, 48, 50–1
Nazism 88
neoliberalism 43, 79
Netanyahu, Binyamin 89
New START treaty 91, 93
New York Times 48–50
Nixon, Richard 27
Non-Proliferation Treaty 26–7, 29, 47, 63, 65
non-violence 70
non-voters 71, 73–4
Nuclear Age 14–16, 31, 33
nuclear weapons: activism against 62–3; in Cold War 30–2; elimination of 29–30, 44–5n7; Iranian 46; popular mobilization against *40*, 42; at Russian border 33–4; small 49; threat of 6, 9–10, 25, 43, 81, 97; and Trump administration 78, 91
nuclear-weapons-free zones 27–9, 46–7, 64, 66
Nunn, Sam 29

Obama, Barack 28, 33, 36, 47–8
Ocasio-Cortez, Alexandria 95
Old South Church, Boston 2, 86
Operation Able Archer 9–10, 31–3
opiate abuse 81–2

Pakistan 21, 27, 65
Palestinians 100–1

Paris climate agreement 21
Perroots, Leonard 32
Perry, William 29–30
Petrov, Stanislav 10, 32
Pollin, Robert 80, 83
Postol, Theodore 92
Pramas, Jason 66
Progressive International 90
Puerto Rico *20*

Rai, Milan 101
Reagan, Ronald 27, 29, 31, 82, 92
Republican Party 8, 21–3, 57, 76
Rumsfeld, Donald 27, 68
Russia: and nuclear weapons 63; US and NATO threat to 31–2, 34, 36–8, 51, 78, 92–3; in World War II 88

Saddam Hussein 27
Sakwa, Richard 37
Sanders, Bernie 59, 72
Saudi Arabia 89, 94
Schultz, George 10
Shawn, Wallace 4–5
Shifrinson, Joshua Itzkowitz 36
social movements 79–80, 84
solidarity 12, 85; international 79–80
Soviet Union 34–5, 37
Spanish Civil War 86–7
Stanislaus National Forest *18*
Sunrise Movement 95, 98
Syria 19

third party 71–3
Trump, Donald 2, 6; and climate change 22–3; and lesser-evil voting 71–2; National Security Strategy 78; and nuclear threats 33; and nuclear treaties 91–3; supporters of 59
Trump-Pompeo-Bolton triumvirate 89, 91
Twain, Mark 87

Ukraine 36
ultranationalism 79, 82, 89–90
United Arab Emirates 89

United Nations 15, 30, 48, 65
United States: cultural backwardness of 42–3; as cultural backwater 56; and fossil fuels 94–5; military policy of 36–7; and Nazi Germany 87–8; and nuclear threat 27–8, 31–3, 47–8, 63–5, 91–2; private interests capturing 9

Varoufakis, Yanis 90
Vietnam War 2, 54–5, 67–8, 100

Wallace Action Fund 12
Wintonick, Peter 101
Working Families Party 73